Rupert Dum_ _ _ _ _

A Sceptic's Guide to the Universe

With help from Janet Bloye and Dorothy Williams
2023

50E

Overview

This essay explores the possibility that electrons and photons are different manifestations of the same particle and that photons gradually lose energy (increase in wavelength) as they travel through space often for billions of years before returning to atomic orbit as an electron. If this is true then an observed increase in wavelength (red shift) does not necessarily indicate a separation of objects in space and suggests that the universe may not be expanding as it is presently thought to be. Read on.

Atoms, electrons and photons

1. **Atoms**

It is generally agreed that all materials are composed of atoms. With the exception of hydrogen, which has no neutron, all atoms consist of a nucleus comprising a combination of protons and neutrons around which electrons are held in orbit by electromagnetic attraction. Protons have a positive electromagnetic charge, electrons have a negative charge and neutrons, which comprise an electron and a proton closely bound together, have no charge.

All atoms have the same number of electrons as they have protons and, with the exception of hydrogen which has no neutron, have as many or more neutrons than they have protons. Iron for example has 26 protons and 26 electrons but 27 neutrons.

Electrons orbit around the nucleus of an atom in distinct shells. The first shell has a capacity of 2 electrons, the second and third shells 8, the fourth and fifth shells 18, and the sixth and seventh 32 electrons. In each case the outermost shell is the most reactive.

With the exception of hydrogen (whose single proton has a mass of 1836 times that of the single electron that orbits around it), the mass of the nuclei of all atoms is at least 3675 times greater than the combined mass of all of the electrons that orbit around it.

An atom is often likened to a diminutive solar system with the nucleus of the atom representing the Sun and the electrons representing the planets. The Sun is approximately 713 times more massive than the combined mass of all the planets and asteroids which orbit around it.

Just as in our solar system, where a few relatively small planets orbit around the Sun in a huge volume of space, even in the densest of materials (such as solid steel) electrons orbit around atomic nuclei in relatively large volumes of space.

Although there are similarities between the arrangement of the solar system and that of a single atom there is, of course, an appreciable difference in scale. Whereas our solar system is thought to be some 287 billion kilometres in diameter, a particle of copper the size of a pinhead may comprise more than a thousand billion atoms.

Note.
1. Although it is thought that protons may be constructed from quarks, since quarks appear to exist only in combination with other quarks there is some doubt as to whether they can be regarded as independent particles. For the purposes of this essay I regard protons and electrons/photons as the smallest independent particles of matter and the basic building blocks of the universe.

2. Electrons and Photons

An electron is a negatively charged particle in orbit around the nucleus of an atom and is generally described as a tiny spinning sphere.

At present it is widely believed that electrons and photons are separate entities with very different properties. An electron is generally regarded as a subatomic particle which has both form and mass, while a photon is generally considered to be a massless and formless quantity of pure energy which exists only when travelling at the 'speed of light'.

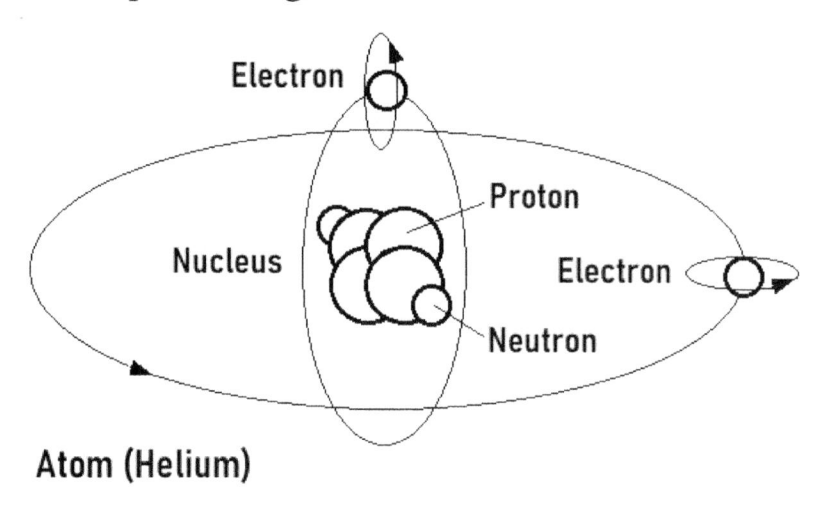

Atom (Helium)

Whilst I agree with the definition of an electron, I find it difficult to accept the idea that a photon is an amorphous quantity of energy which has no mass and exists only when travelling at exactly 299,792,458 metres per second in a perfect vacuum (which I understand does not exist anywhere in the universe).

This essay explores the possibility that

that electrons and photons are simply different manifestations of the same particle and examines the effect that this might have on our understanding of the universe.

Note

1. The absolute maximum speed that anything can travel in the universe (299,792,458 m/s) is often referred to as c or 'the speed of light'.

The actual speed of light

1. The Jefferson Laboratory experiment

I do not doubt that there is an absolute maximum speed for anything that travels in the universe. It is generally agreed that this speed should be taken as exactly 299,792,458 metres per second.

Experiments carried out at the Jefferson Laboratory in Virginia found that, electrons (which have a tangential velocity of approximately 2,180,000 m/s in atomic orbit) could, with a boost of energy, be induced to leave atomic orbit and travel in a generally straight line. It was also noticed that the speed of the electron in free flight depended on the amount of energy that it was given when it left atomic orbit.

It was found that, although only a moderate amount of additional energy was required to accelerate an electron from the point at which it left atomic orbit up to a velocity of 90% of c, the amount of energy required to accelerate it closer to the maximum possible speed c (299,792,458 m/s) began to rise exponentially so that an enormous amount of energy was required to increase its speed from 95% to 99.9% of c.

Nonetheless scientists at the Jefferson Laboratory, using a colossal amount of additional energy, managed to accelerate the electron to a speed of 99.9999992% of c before the experiment was terminated.

The fact that the Jefferson Laboratory, despite the use of a colossal amount of energy, failed to accelerate an electron to c, suggests that it may not be possible for any particle that has mass to achieve and sustain the theoretical maximum speed of 299,792,458 m/s regardless of the amount of energy applied.

Here we have to pause for thought. If the particle that the laboratory was accelerating was no longer in orbit around an atom and was carrying energy in a generally straight line at a speed closely approaching the theoretical maximum speed c, then surely it was behaving more like a photon than an electron and suggests that an electron may simply assume the characteristics of a photon when it is given sufficient energy to leave atomic orbit.

If this is true then photons, like electrons, have mass and therefore can approach but not sustain the theoretical maximum possible speed of 299,792,458 metres per second.

If we accept the idea that photons are in fact free electrons then we have to consider the possibility that they travel at different speeds and wavelengths depending on their energy levels.

From the graph (below) it would seem that a higher energy photon (such as a gamma ray) may travel at around 99.9% of c while a lower energy photon (such as an extremely low frequency radio wave) may travel at as little as 95% of c.

Speed (% 299,792,458)

```
0    10   20   30   40   50   60   70   80   90  100%
```

Gamma Rays

Visible Light

Electron in atomic orbit
2,180,000 m/s

Low frequency radio waves

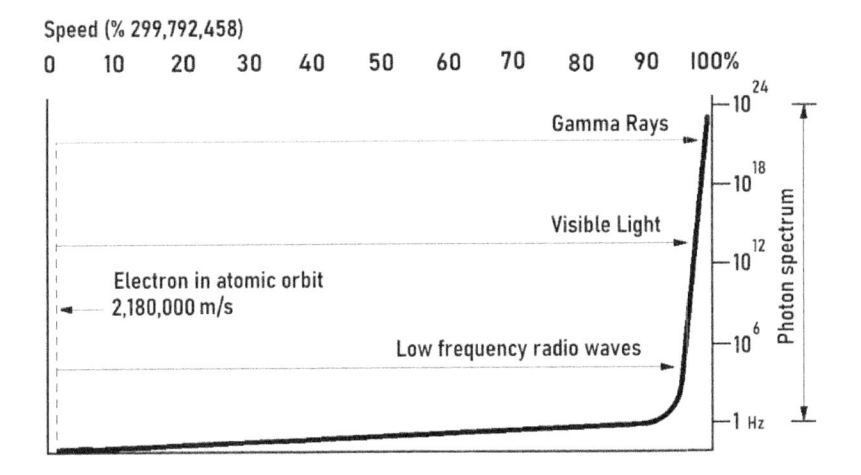

10^{24}
10^{18}
10^{12}
10^{6}
1 Hz

Photon spectrum

Increase in energy requirement as photon approaches maximum theoretical speed c (299,792,458 m/s)

From the Jefferson Laboratory experiment it would seem possible that an electron, when given sufficient energy to leave atomic orbit, simply changes its role from a captive atomic particle to a free particle which is capable of carrying energy.

Under these circumstances a photon, rather than travelling continuously at the maximum theoretical speed c, would begin its journey at the velocity given to it when it left atomic orbit, then gradually lose energy as it collides with (or is influenced by) other small particles of matter during its long journey through space until it has so little remaining energy that it is captured by an electron-depleted atom somewhere in the universe.

If this is the case then an extremely high frequency photon (such as a gamma ray) radiated from a distant star could, after travelling through the universe for billions of years, arrive at the Earth in the form of a low frequency electromagnetic wave (such as a long wavelength radio wave).

What might at first have appeared to be an indecipherable radio signal from a distant galaxy may in reality be the remnants of gamma rays which have lost most of their energy during a very long journey through space. It follows from this that a photon of lower frequency radiated from the same galaxy would simply have run out of energy and been absorbed by an electron-depleted atom before reaching the Earth.

Notes

1. One idea that emerges from this discussion is that photons, like everything else in nature, decay with time. This infers that the initial infusion of energy given to a photon when it leaves atomic orbit will (unless in collision with matter) be slowly exhausted as it travels through space for billions of years.

2. According to my calculations using the equation $E=mc^2$, where m is 9.109×10^{-31} kg (the mass of an electron) and c is 299,792,458 m/s, the amount of energy (E) required to accelerate a photon as close as possible to the theoretical maximum possible speed c, is five hundred and eleven thousand electron volts (511 keV).

How many photons are there in the universe?

The amount of photons generated by man-made devices (such as street lights) is negligible in comparison to the quantity of photons radiated by a star.

It has been calculated that our sun radiates four million tonnes of photons every second. Since our sun is an average sized star and there are more than one hundred billion stars in our galaxy (the Milky Way), it can be seen that more than four hundred million billion tonnes of photons are radiated into space, by the stars in our galaxy alone, every second. What we call space is by no means empty nor is it a perfect vacuum.

Red shift

1. **The Doppler effect**

You will almost certainly have noticed that the tone of the siren of an emergency vehicle (which emits sound at a constant frequency), appears to change pitch as it passes a stationary observer. The tone of the siren seems to increase in frequency as the vehicle approaches the observer then reduce in frequency as the vehicle moves away again. This is the Doppler effect.

A similar effect can be detected when the frequency of light (or any other photon), radiated from an object in space, increases or decreases as the object approaches or moves away from the observer.

If the frequency of a photon were to remain absolutely constant over a period of billions of years then the Doppler effect (red shift) would give a reliable indication of the rate of separation between an object and the observer but if, as has been suggested, the frequency of a photon (like everything else in the natural world) decays with time then blue light will naturally decay towards red light and the presence of 'red shift' will not necessarily indicate separation between the object and the observer.

Since 'red shift' is at present widely used to estimate the age, size and rate of expansion of the universe we will have to take a closer look at this aspect of the problem later in the essay.

Natural decay

I understand that, during the early years of the twentieth century, astronomers set out to determine the size and age of the universe. They did this by first identifying the farthest star that could be observed from the Earth then, using the 'speed of light' and measurements of 'red shift', eventually decided that the universe was 13.799 billion years old and was expanding rapidly.

In those days it was assumed that all electromagnetic waves travelled at the 'speed of light' and that 'red shift' was a reliable way to determine the rate of separation between objects in space.

It was also thought that since all electromagnetic waves travelled endlessly at 'the speed of light' there was no limit to the distance that a photon could travel.

In this essay it is assumed that photons will begin to lose energy at the instant that they are driven out of atomic orbit and will continue to lose energy (decelerate) over a period of billions of years until they are exhausted to the point where they are able to rejoin the orbit of an electron depleted atom.

The first phase of deceleration is usually referred to as the electromagnetic spectrum and comprises photons ranging from ultra high energy gamma rays travelling at a speed very close to the maximum theoretical speed c down to ultra low energy radio waves travelling at around 95% of c. (See appendices A and C).

In this phase of deceleration gamma rays, for example, will slowly degenerate into X-rays and blue light into red.

If this is true then 'red shift' will occur naturally as the frequency of a photon decreases without indicating separation between objects in space and therefore the widely held view that the universe is continuously expanding will have to be reconsidered.

If photons do lose energy with time then it follows that there must also be a limit to the distance that a photon can travel before it runs out of energy and returns to atomic orbit.

The rate of deceleration of photons within the electromagnetic spectrum may be estimated by examining the *most distant* object that can be seen from the Earth. If the image of the distant object arrives at the observatory on Earth in the form of visible light then it is reasonable to assume that it was radiated from the distant object in the form of much higher energy (invisible) photons some billions of years earlier. The distance of the object from the Earth is then equivalent to the time that it takes for the higher energy photons to degenerate into visible light.

When the speed of a photon falls below that of an ultra low energy radio wave (i.e. below the range of the electro magnetic spectrum) it will continue to decelerate for billions years at an ever decreasing rate until it is exhausted to the point where it rejoins the orbit of an electron depleted atom somewhere in the universe.

Notes

1. It would seem that photon decay (a very

gradual reduction in speed and increase in wavelength) is caused by minute quantities of atoms (typically one atom of hydrogen per cubic centimetre) which occur even in intergalactic space, thereby preventing the formation of a perfect vacuum and ensuring that a photon cannot maintain the maximum theoretical speed of 299,792,458 m/s.

2. *If, as proposed in this essay, photons gradually lose energy (and increase in wavelength) as they travel through space for billions of years, then a high frequency photon, such as a gamma ray, may arrive at an observatory on Earth in the form of an x-ray, visible light or a radio wave depending on the distance that it has travelled.*

3. *Galaxy HD1, which lies some 13.463 billion light years away from our solar system, is one of the most distant objects that can be observed from the Earth in the form of visible light. If, as explained above, photons gradually lose energy as they travel for immense distances through space, then photons that arrive at the Earth from HD1, in the form of visible light, must have been*

radiated from within the HD1 galaxy in the form of high frequency (gamma) rays.

4. It follows from this that if, during a journey of 13.463 billion light years, photons will decelerate at an an ever increasing rate from ultra high frequency gamma rays (which lie at the head of the spectrum) to visible light (which lies towards the centre of the spectrum) then the length of the entire electromagnetic spectrum may be in the region of thirty billion light years.

5. As the distance between an object in space and an observer on Earth approaches the absolute maximum distance that a photon can travel through the electromagnetic spectrum, then reception at the observatory will be reduced to a few ultra low frequency radio signals shortly before all contact is lost.

6. A light years is a measure of distance equivalent to 9.46 trillion kilometres.

7. Throughout this essay the expressions 'electromagnetic spectrum' and 'photon spectrum' are used interchangeably.

Photon journey times

In this essay it is proposed that the electromagnetic spectrum comprises photons with speeds ranging from ultra high frequency gamma rays down to ultra low frequency radio waves.

Whereas in deep space the difference in journey times between high and low frequency photons is measured in terms of billions of years (and is of crucial importance to our understanding of the size and rate of expansion of the universe), on Earth the difference in journey times between high and low frequency photons is measured in fractions of a second and is too small to be of any real significance.

On Earth, for example, the difference in journey times between an ultra high frequency gamma ray (travelling at 99.9% c) and an ultra low frequency radio wave (travelling at 95% c) when travelling once around the equator is less than one hundredth of a second.

Notes

1. The equator is some 40,075,017 metres in circumference.

2. It is possible that the extremely small difference in journey times between electromagnetic waves of widely different frequencies, when measured on Earth, may have encouraged early scientists to accept the proposition that all electromagnetic waves travel at the same speed.

Frequency, wavelength and the electromagnetic spectrum

1. Frequency and Wavelength

The frequency of a photon is the number of times that it rotates in one second and is measured in cycles per second (Hertz). The wavelength is the distance that the photon travels while it completes one rotation and is measured in metres.

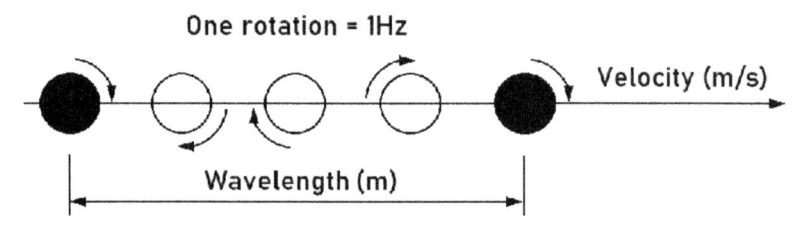

Photon

The relationship between the frequency and wavelength of a photon can be derived from the equation:

Wavelength(m) x Frequency(Hz) = Velocity(m/s)

The frequency of rotation of a photon will depend on the amount of energy infused into it at the point of transmission. The more energy infused into the photon as it leaves atomic orbit the greater its frequency and the shorter its wavelength.

Whereas a radio wave may have a frequency as low as three Hertz and will be easily reflected by a thin sheet of aluminium foil, a gamma ray, which may have a frequency of more than thirty thousand billion billion Hertz, is able to penetrate several centimetres of solid steel.

2. **The Electromagnetic Spectrum**

The electromagnetic spectrum includes photons with frequencies ranging from less than three cycles per second (an extremely low frequency radio wave) to more than a thousand billion billion cycles per second (a high frequency gamma ray).

Visible light occupies a narrow band of wavelengths near the centre of the spectrum ranging from 400 to 700 nanometres (400 to 700 billionths of a metre).

Photons passing through space are normally unpolarised (i.e. the planes of rotation of the spinning photons are randomly distributed). Polarisation is said to occur when the plane of rotation of all photons in a particular stream have the same alignment (e.g. in a horizontal or vertical plane). Some high frequency radio waves are intentionally polarised in order to reduce the transmission power requirement or to improve received signal strength.

Visible light can be polarised by using a polaroid filter which allows photons of only one alignment to pass through it. Photons of other alignments are absorbed or reflected by the filter so that the total amount of light passing through the filter is reduced proportionately.

Electricity, magnetism and gravity

It is generally accepted that electricity, magnetism and gravity are similar in some respects to electromagnetic forces but are not included in the electromagnetic spectrum.

1. Electricity

Electricity is usually described as the movement of electrons through a conductor (such as a copper wire).

When an electrical charge (electromotive force) is applied to one end of the conductor, a stream of electrons, or pulse of energy, travels along the conductor at around 90% of c and emerges as a stream, or pulse, of electrons at the far end of the conductor.

Electricity has been described as an electromagnetic wave guided by, but flowing overwhelmingly outside of, the conductor. This suggests that a stream or pulse of electrons, rather than travelling through a dense material (such as a copper wire), will take the line of least resistance and travel through a less dense surrounding medium if it is available.

Electricity may be supplied as a direct or alternating current. In a direct current the stream of electrons is unidirectional. This does not necessarily mean that a single electron will pass from one end of the conductor to the other. Some people believe that an electron entering at one end of the conductor will displace an electron from the next atom and so on until an electron emerges at the far end of the conductor. (The effect is similar to the way that the impact from the ball at one end of a Newton's cradle transfers the energy instantaneously to the ball at the other end of the cradle).

Where electricity is supplied in the form of an alternating current, the direction of a group of electrons, or pulse of energy, travelling along (or guided by) the conductor is reversed fifty or sixty times a second (50 or 60 Hz) so that pulses of useful energy arrive at the far end of the conductor every fiftieth or sixtieth of a second.

Although electrons, when travelling in the form of electricity, may travel outside the conductor, they remain in close association with atoms in the conducting material and are therefore not generally regarded as photons (electromagnetic waves).

Perhaps the easiest way to create a stream of photons (such as visible light) from a stream of electrons (electricity) is to pass it through a light emitting diode (LED). A light emitting diode is in effect a simple electron to photon converter.

Light can be converted back into electricity using a photovoltaic (silicon) cell.

2. Magnetism

Electrons carry a small magnetic charge. The plane of rotation of the electrons in a magnetic material, such as iron (Fe), nickel (Ni) or cobalt (Co), are normally randomly distributed and the small magnetic force carried by electrons in individual atoms tend to cancel each other out so that there is little or no magnetic field. In a permanent magnet (as the name implies) the planes of rotation of the electrons are permanently aligned and a strong magnetic field is evident.

A temporary magnetic field can be induced in a bar of a magnetic material (such as soft iron) by sending an electrical current through a wire coiled around the bar. This causes the electrons to align themselves and creates a magnetic force which will remain in place until the electrical current is switched off when the electrons revert to random alignment and the magnetic field disappears.

Whether the magnet is permanent or temporary the magnetic force is strongest at each end of the bar (forming 'north' and 'south' poles).

3. Gravity

The force of gravity is not fully understood but it has been described as a relatively weak force that travels at 'the speed of light'. With so little information to start with, it is not easy to describe the mechanism of gravity but there are a few clues.

In our solar system the strongest gravitational force is exerted by the Sun which holds eight planets, and an incalculable number of smaller particles of matter, in orbit.

It would seem that the smallest unit of gravity is created when an atom is deprived of an electron. Under normal circumstances each atom has the same number of (positively charged) protons as it has (negatively charged) electrons and the two opposing charges cancel each other out but when an atom is deprived of an electron (for example when it is radiated into space in the form of a photon), the small attractive force of the remaining proton will try to draw matter from elsewhere. This small attractive force is, I think, the building block of gravity.

If this is the case then our sun, which has been radiating huge quantities of photons into space for more than four billion years, will have built up, within its core, a colossal mass of dense electron-depleted atoms (cations) which at present possess sufficient gravitational force to hold several planets and an enormous amount of smaller matter in orbit.

On a very much smaller scale the planets themselves also tend to accumulate dense, positively charged matter in their cores. Jupiter for example, which has only 0.09% of the mass of the Sun, still possesses sufficient gravitational force to hold seventy nine natural satellites in orbit.

Stars

Young stars are huge spheres of (predominantly) hydrogen in space.

Due to the size of even the smallest of young stars, extreme temperature and pressure deep within their cores result in protons and electrons being fused together by nuclear energy to form neutrons which enable large quantities of helium to be produced before smaller quantities of more complex atoms are created.

Extreme temperatures and pressures deep within the core of the star will also infuse electrons with sufficient (nuclear) energy to cause them to leave atomic orbit and carry energy out into space in the form of photons of all frequencies (including visible light).

Where the energy generated by nuclear fusion within the core of a star is balanced by photons radiating energy into space, the star may remain in a stable condition for billions of years. Our sun has managed to maintain this balance of energy for more than four billion years.

But where a star produces more nuclear energy than it can dispose of by radiation, the temperatures and pressures within its core may increase to the point where they trigger a chain reaction leading to a devastating nuclear explosion (a supernova) which reduces the star and everything within its vicinity to a cloud of small particles (a nebula).

If, on the other hand, *insufficient* nuclear energy is created to maintain the balance of energy then the gases (mainly hydrogen and helium) of which the star is composed, may begin to cool then shrink dramatically causing the star to collapse in on itself to form a very much smaller but extremely dense neutron star.

Notes

1. Our sun is about 1.4 million kilometres in diameter and has a mass of some 330,000 times that of the Earth. Deep in its core hydrogen, at a temperature of in excess of fifteen million degrees Celsius and a density of ten times that of lead, is converted by nuclear energy into helium at a rate of 600 million tonnes per second. Nuclear activity within the core of the Sun also provides sufficient energy to radiate four million tonnes of photons into space every second. It would seem that radiation of photons into space is the means by which the Sun disposes of surplus nuclear energy.

2. In view of the enormous size and mass of stars in comparison to the planets which orbit around them, it is almost certain that the vast majority of exhausted photons are absorbed by any one of the two hundred billion trillion stars in the universe. In this way it seems that stars act as recharging stations for photons which have lost their energy after travelling through the universe for billions of years.

The Big Bang and a cloud of subatomic particles

1.   **The Big Bang**

It is at present widely believed that all the matter in the universe was created almost instantaneously from nothing 13.799 billion of years ago.

This curiously unscientific (almost biblical) theory is known as the Big Bang:

> *In the Beginning there was a Big Bang and everything in the Earth and in the Heavens was created from Nothing in the blink of an Eye.*

The Big Bang theory breaches just about every law of physics that there is (including the one which states that matter cannot be created or destroyed).

It's up to you whether you decide to believe that all the matter required to make the universe was created from nothing in a fraction of a second 13.799 billion of years ago, or just settle for the idea that all of the matter required to form the universe has always been in existence.

If you decide that all the necessary matter was already in existence then all you have to do is to decide in what form the matter may have existed immediately before the universe, as we know it, began to take shape.

Of the many alternatives that have been proposed, I will try to summarise two theories which comply with the laws of physics.

2. **A Cloud of Subatomic Particles**

One credible theory is that the universe developed from a cloud of subatomic particles.

Imagine that, long ago, there was a cloud of protons and photons evenly distributed in *the same volume of space, and having the same total mass,* as the present universe.

At some time in the distant past, a (negatively charged) photon in one region of the cloud, was drawn into orbit around a (positively charged) proton to form a partnership which we now call an atom of hydrogen. Then other hydrogen atoms began to form at various locations within the cloud and this procedure continued until vast quantities of hydrogen were formed.

Inevitably one of the orbiting photons was drawn into direct contact with a proton to form a (neutrally charged) neutron.

With the development of the neutron, atoms of helium (which comprise two protons, two electrons and two neutrons) began to form. For a long time only hydrogen and helium were created before much smaller quantities of more complex atoms began to develop.

In time these atoms gathered into groups which became larger and denser until sizeable accumulations of matter began to develop. Some accumulations of matter developed to the size of planets. Larger accumulations became stars and began to radiate photons back into space. Billions of stars formed into galaxies and billions of galaxies into a universe.

What began as a fine cloud of subatomic particles (a nebula) was gradually transformed into widely spaced accumulations of dense matter which occupied *the same volume of space and had the same total mass* as the particles in the original cloud.

In this way matter was neither created nor destroyed and the laws of physics were obeyed.

Notes:

1. Trying to define where and when the first atom of hydrogen may have been created within a cloud of subatomic particles is rather like scattering seeds in a field and trying to predict which seed will germinate first. The creation of the first atom of hydrogen within the cloud may have occurred anywhere and at any time. In this proposition the next hydrogen atom would be created somewhere else in the cloud and this haphazard pattern of development would continue until atoms of hydrogen began to form into clusters and the universe, as we know it, began to take shape.

2. It is worth noting that hydrogen and helium were the first two elements to develop and still remain by far the most abundant elements in the universe. In our solar system the Sun (which came into being some 4.6 billion years ago) is composed of 73.4% hydrogen and 25% helium. The two largest planets in our solar system (Jupiter and Saturn) are also composed mainly of hydrogen and helium.

3. The suggestion that the first hydrogen atoms may have formed at various different locations within the cloud gives credence to a commonly held belief that the universe began to form everywhere at once.

4. One interesting aspect about the haphazard development of matter within a cloud of particles is that its growth pattern is very different to that of an explosion. An explosion occurs at a particular time and

location and tends to expand spherically.

5. Had the universe been created from nothing in a fraction of a second then the exact time at which it was created could be calculated with a degree of accuracy but if the first atom of hydrogen was formed at random in a cloud of basic particles then the exact time that the process of change began cannot be determined with any degree accuracy.

3. Endless Regeneration

In this theory the explosion of a massive black hole at the centre of a galaxy reduced the whole galaxy to a huge cloud of small particles.

Over a period of many millions of years protons, which had been widely scattered by the explosion, began to capture low energy photons from the vast numbers of photons, which were always streaming through the universe, and formed atoms of hydrogen. Gradually helium then smaller quantities of more complex atoms began to develop which, together with other small particles that had managed to survive the explosion, began to form sizeable accumulations of matter.

Over a period of billions of years some of these accumulations of matter grew to the size of planets. Some became large enough to form into stars and began to radiate photons back into space.

As photons were radiated into space, positively charged protons which remained in the core of the star began to develop sufficient gravitational force to attract small particles of matter from the surrounding area and the star continued to grow.

As a new star grew it began to develop sufficient gravitational force to draw planets and other smaller accumulations of matter into its orbit to form a solar system. Simultaneously billions of other solar systems began to develop within the debris cloud and together gradually formed a new galaxy.

At the centre of the growing galaxy, where gravitational forces were at their strongest, one enormous old star grew so massive that it developed sufficient gravitational force to draw whole planetary systems into itself to form a colossal and extremely dense invisible mass known as a black hole from which nothing (including radiation) could escape. Eventually the black hole became so massive that it triggered a nuclear explosion of such violence that the whole galaxy was once again reduced to a cloud of small particles and the whole process of development began again.

Matter was neither created or destroyed and the laws of physics remained intact.

Notes

1. Black holes are invisible because they no longer radiate light (or any other electromagnetic wave) and absorb any electromagnetic wave that approaches them.

2. The black hole (Sagittarius A) which lies at the centre of our galaxy is thought to be at least a million times more massive than our sun.

3. It's possible that our galaxy (the Milky Way) began to develop from the cloud of small particles left behind by the explosion of a massive black hole (a Big Bang) which, 13.77 billion years ago, destroyed one of the millions of galaxies which had already been in existence for billions of years.

The universe by observation and what lies beyond

1. The Universe by Observation

Almost all of our information about the universe is gained by observation from the Earth but, as has already been suggested, observations of remote sources of radiation from a single point in space can be very misleading.

Photons radiated from distant stars may have taken billions years to reach the Earth during which time once brilliant stars have faded. Some have risen to prominence and others have disintegrated. Nothing has remained in the same place.

Even the sequence of events as observed from the Earth is confusing. If the image of a remote object has already passed through a part of the universe which is later occupied by a closer object, then the image of the more distant object will, when viewed from the Earth, appear to be in front of the closer one.

In this way even the sequence of images observed from the Earth is not a true representation of the events which have occurred in the universe. What we see when we look at the night sky is a scrambled arrangement of images of things which have long since moved, changed or ceased to exist.

2. The Size of the Observable Universe

Although it is not possible to determine the size of the *whole* universe, attempts have been made to estimate the size of the *observable* universe by studying changes in the frequency of photons received from the most distant parts of the universe that can be observed from the Earth.

At present it is estimated that the radius of the observable universe may have expanded to 46.5 billion light years but, as already explained, this figure is subject to review.

If, as has been suggested, photons like everything else in nature decay (lose energy) with time, then 'red shift' (an increase in wavelength) will occur when there is no separation between any two objects in space.

If this is the case then any calculation relying on 'red shift' to determine the size and rate of expansion of the universe will be unreliable and the universe may not be expanding as it is presently thought to be, if at all.

3.   **What lies Beyond**

What lies beyond the edge of the observable universe is of course unknown but there is room for speculation.

If the universe was formed from only part of an infinitely large cloud of subatomic particles then, unless the remainder of the cloud is still lying dormant waiting for the first hydrogen atom to appear, it is probable that large numbers of universes have already developed elsewhere in the cloud.

If on the other hand a single infinitely large universe exists then the universe, by definition, has no bounds.

We are left only with the question of why we cannot detect anything beyond the limit of the observable universe. In this essay it is suggested that our ability to detect matter in the outer regions of the universe is limited by the maximum distance that a photon, even when provided with the maximum possible amount of energy, can travel through space before it decays to the point where it has so little remaining energy that it is once again absorbed into atomic orbit. this is true then it is possible that we may never be able to detect anything which lies beyond the maximum distance that an extremely high energy photon can travel before it runs out of energy.

Appendix A: Photon spectrum

	Wavelength	Frequency	Speed
			100%c
		10^{18} Hz	
Gamma rays	10^{-12} m		
X-rays	10^{-9} m	10^{15} Hz	99%c
Visible light Infrared	10^{-6} m	10^{12} Hz	
			98%c
Microwave	10^{-3} m		
		10^{9} Hz	
			97%c
UHF	1 m		
		10^{6} Hz	
Radio waves	10^{3} m		
			96%c
		10^{3} Hz	
ELF	10^{6} m		
		1 Hz	95%c

Appendix B: Energy requirement

Speed (% 299,792,458 m/s)

0 10 20 30 40 50 60 70 80 90 100%c

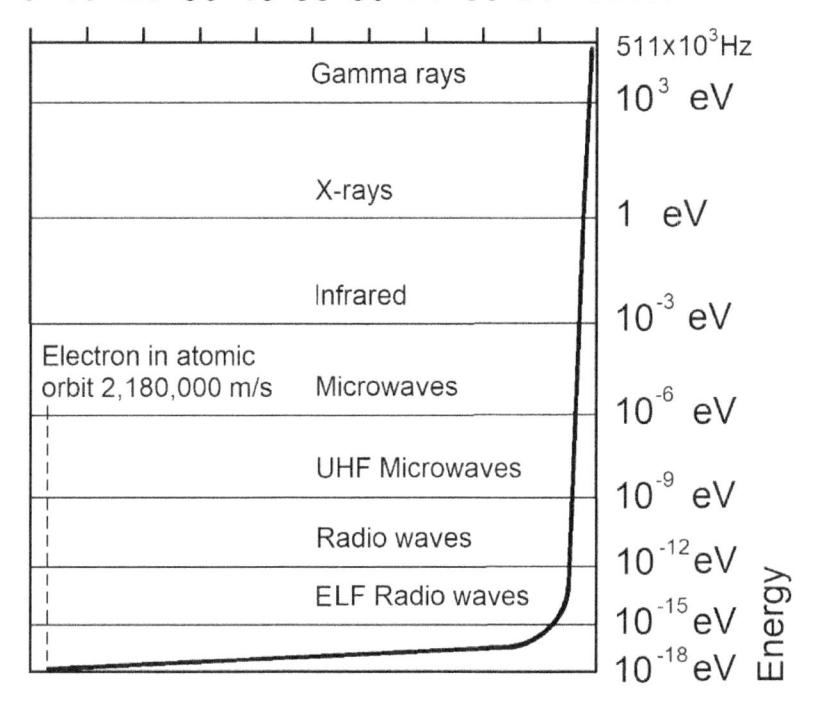

Gamma rays	511×10^3 Hz
	10^3 eV
X-rays	1 eV
Infrared	10^{-3} eV
Electron in atomic orbit 2,180,000 m/s — Microwaves	10^{-6} eV
UHF Microwaves	10^{-9} eV
Radio waves	10^{-12} eV
ELF Radio waves	10^{-15} eV
	10^{-18} eV

Energy

Increase in energy requirement as photon approaches the maximum theoretical speed (299,792,458 m/s)

Appendix C: Wavelength and frequency below photon spectrum

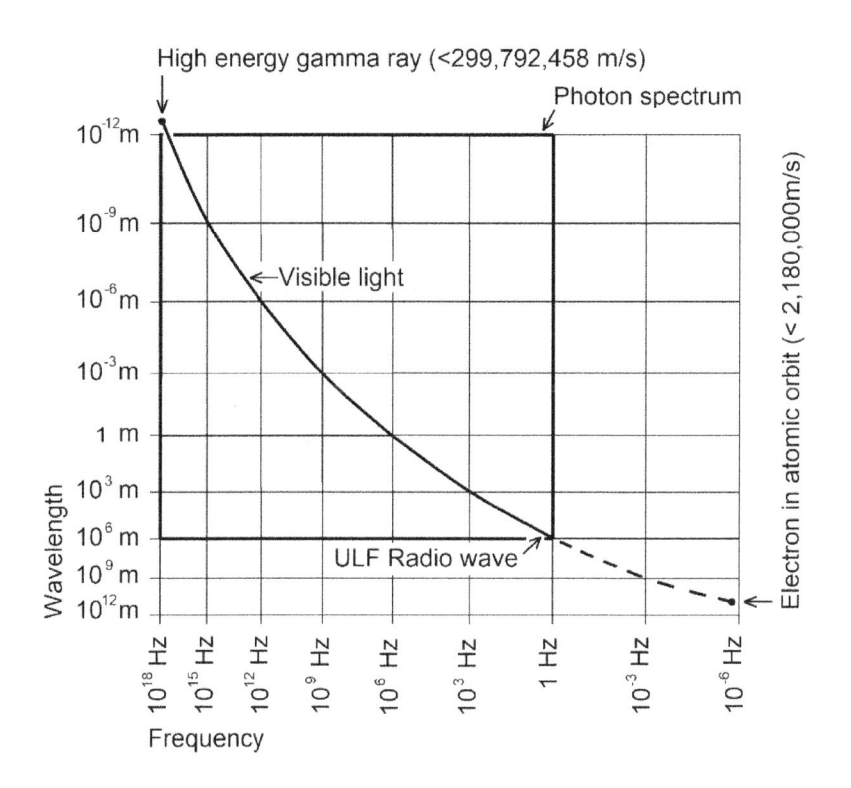

Printed in Great Britain
by Amazon